Follow along and you will soon see
just what makes a fact family.

Let's start right away. You can have fun.
Let's make a fact family with
three, two, and one.

There are four equations
in a fact family. The first is
two plus one is equal to three!

Now here's the second equation to do.

Three minus one is equal to two!

$$3-1=2$$

Take another look and see,
one plus two is equal to three.

A final equation, and we're done—
three minus two is equal to one.

Four equations are the key to creating any fact family!

$2+1=3$ $3-1=2$ $1+2=3$ $3-2=1$

A new fact family of **five, three, and two.**
These fish present an equation for you.

Five minus three equals two,
when you subtract.
Can you name another equation
to get a new fact?

$$5 - 3 = 2$$

There's no time to lose,
no need to wait.
Here's a fact from the family
three, five, and eight.

What's the opposite fact
for the equation you see?
What's another addition fact?
What might that be?

$3+5=8$

Take two numbers exactly the same.
Add and subtract for a Fact Family Game.

Six plus six is twelve dogs in a row.
What's the opposite equation?
Do you know?

6 + 6 = 12

Twelve balloons plus eight, in the sky.

$12 + 8 = 20$ $20 - 8 = 12$

Name the facts for this family. Give it a try!

$8 + 12 = 20$ $20 - 12 = 8$

Now you know about fact families.
Add and subtract—make a fact family tree!